Newly emerged Glanville Fritillary,
Isle of Wight.

BUTTERFLY SAFARI

'Butterflies are the souls of summer hours and here
we get to gaze upon those souls – an epic achievement
that will inspire young and old to pay closer attention
to the miraculous lives all around us.'
Patrick Barkham, author of *The Butterfly Isles*

Female Chalkhill Blue in flight,
Nottingham Hill, Gloucestershire.

BUTTERFLY SAFARI

Andrew Fusek Peters

BIRD EYE BOOKS

CONTENTS

Front cover: Adonis Blue, Mill Hill, Sussex.
Left: Swallowtail flight sequence, the Doctor's Garden, Strumpshaw Fen, Norfolk.

INTRODUCTION

Photo by Polly Peters.

I want to thank my two fellow Aurelians Craig Jones and David Williams, who took it upon themselves to drive me all over the UK and fill in the gaps of my list and share their butterfly experience and wisdom. This book is dedicated to them.

Left: Female Common Blue at sunset, The Bog, Stiperstones.

My love affair with butterflies began in 2018. The Painted Ladies were about their work in our garden and I suddenly wondered, could I dare to try and capture a butterfly in flight? Just articulating the question seemed outrageous and I could hear my camera kit complaining at such impossibility. But study, perseverance and a silly number of hours finally enabled me to catch flight alongside aerial take-off sequences that showed butterflies doing what they do best.

As I grew to understand butterflies, I desired to get nearer. Macro is such a dull word that can in no way reflect what happens close-up, yet its meaning – both long and large – sums up the effort spent making a tiny world grow in scale until we can see minute detail with the naked eye. There are worlds within worlds and a butterfly can be dressed in up to a million scales – hence their name from the Greek, *lepidoptera*, meaning scaly wings. Such micro-architecture is breath-taking, as if evolution went hand-in-hand with the pagan spirit of nature that decreed both sun and blue sky be reflected in layers of tiny scales, hanging on hooks like an oak shingle roof.

Back in the 17th century, John Ray put it perfectly: 'What is the use of butterflies? To adorn the world and delight the eyes: to brighten the countryside like so many golden jewels. To contemplate their exquisite beauty is to experience the truest pleasure.'

But beauty in itself is not enough, as many of our UK butterflies are in trouble. However, there is a rising band of conservationists who use science and sheer hard and mostly voluntary work to study habitat and restore intimate landscapes to suit the particular requirements of individual species. Such efforts in many cases are paying off. The fluttering Wood White that patrols the rides of Bury Ditches high above our village in Shropshire has spread over the hills to Clunton Coppice. In Norfolk, the range of the exotic Swallowtail is contracting, but thanks to conservation work, it is found in greater numbers in its strongholds.

There are gangs of butterfly lovers out there who descend on butterfly sites to dig and cut and manage woods, rides and meadows to make just the right adjustments so that butterflies may mate and lay eggs and the whole cycle of colour and creation begins again.

Where do I fit into all this? My type of digging is pictorial, my conservation work aiming to show through sheer wonder what we could lose if we are not careful. This is no longer the time of vast swishing nets or the killing jar or the pins to spread wings and fix them for the male gaze. The modern Aurelian, the up-to-date butterfly collector, must have concern at his or her heart.

And so to *Butterfly Safari*. It has been an expedition, a four-year dip into our UK species, from the scales that feather and line an antenna to the grandeur of split-second gravity-defying ascension. Here on these pages lies hope and warning. If we are to keep such iridescence and brevity, then we must take good care before it is too late.

Andrew Fusek Peters

Above: Small Tortoiseshell at Stokesay Flowers, Shropshire.
Right: Chequered Skipper flight sequence, Fineshade Wood, Northamptonshire.

APRIL

ORANGE-TIP

The heat on this April day is the echo of the summer some long way off, repeating down the high drovers' road above our village. Forget-me-not has responded, a blue song proliferating in a green shade. On these small high lanes, the only traffic jam is a cacophony of bird song. I am desperate for the first butterfly that turns winter to spring. There! The flash of out-there seventies orange as the male Orange-tip patrols his territory. His colour is both beauty and an early season warning to the birds: 'Don't eat me! I am most unpalatable!' The name *Anthocharis cardamines* tells us that this butterfly is linked to its food plant, lady's smock or garlic mustard.

I have a sudden revelation: nature is here and now, under my very nose! I am happy to be excited by the common, the local, the unexpected and vibrant gift that drifts up and down these ancient lanes in search of love and the urge to new life.

The cattle high on Oakley Mynd walked these droving paths at two miles an hour, right past the milestone that says 'London, 185 miles'. But butterflies are faster, unexpected, never taking off from nectaring in the direction my camera urges. A noble calling then, to catch that intensity of colour, that wing lift, that sudden aerial agitation. What a place to be. My very own nature reserve on my doorstep! For now, it is me and the Orange-tips lifting off the rich blue forget-me-nots, the heat clinging like a coat, the verges bursting with vibrance, the steeple and houses of my village cupped in the hazy valley far below and the glory of a perfect spring day. If this is a beginning, then with gladness in my heart I think that I shall carry on.

Above: Male Orange-tip on dandelion.

Left: Male Orange-tip flying from forget-me-not, south Shropshire.

PEACOCK

The great entomologist James Petiver originally called this butterfly the 'Peacock's Eye', and it's the eyes of the Peacock that always impress me. There could be no better warning to predators than this fierce, staring creature. Their underwing when perching is almost black and the ideal camouflage. The Peacock is a butterfly that hibernates during winter and while walking at Walcot Wood the bluebells give a perfect backdrop to a very worn specimen that has got through the season to now emerge and breed. The Latin name, *Aglais io,* refers to Io, the priestess who served Hera. Hera's husband Zeus fell in love with Io and made love to her while disguised as a cloud. When Hera found out, she turned Io into a cow and had the 100-eyed Argus guard her. Eventually, the myth relates that Hera took the eyes of Argus and put them on the tail of her favourite bird, the peacock. The Peacock has increased its range and can now be found all over the UK.

Right: In mid-August, the next generation of Peacocks are on the wing at Venus Pool and nectaring on lucerne.
Left: I found a roosting Orange-tip on a cloudy day when it would not fly, so was able to use a macro setup to show its remarkable compound eye.
Page 14: A tatty winter survivor flies off bluebells, Walcot Wood, Shropshire.
Page 15: Peacock at Venus Pool, Shropshire.

DINGY SKIPPER AND GRIZZLED SKIPPER

On a good day in late April, the Dingy Skipper and the Grizzled Skipper can both be seen flying together among the spoil heaps and old workings of Llanymynech Rocks. This area was mined for over 2000 years and the great cliff face of the quarry rises from the low-lying fields of the Morda and Vyrnwy rivers. It was designated a nature reserve in 1972 and as it straddles the border I have no idea whether these butterflies are Welsh or English!

I wish the old name of 'Handley's Brown Hog Butterfly' had stuck. Dingy does not seem fair, especially as I kneel in the grass to capture a macro of its wings. If we must go with beige and brown, here are 10 different subtle shades, though admittedly as specimens grow worn and lose their scales they might look a little lacklustre, possibly, dare I say it, dingy.

The Grizzled Skipper is altogether more elusive, especially in flight, when it is impossible to track by eye. It takes two years of visits and lots of other butterfly enthusiasts with sharp eyes to help me capture that checkerboard patterning on the wings as it zips low to the ground. This butterfly belongs to the genus *Pyrgus*, which is from the Greek and means a battlement, an accurate allusion to its highly patterned wing margin. Sadly, since 1976 the Dingy Skipper has lost 61% of its range and the Grizzled Skipper 53%, however, both species are doing well at Llanymynech.

Above: At Prestbury Hill, a Dingy Skipper has a rest on my shoe.
Left: Seen at Llanymynech Rocks, Grizzled Skippers are small and immensely hard to capture in flight.

Above: Grizzled Skipper at Llanymynech Rocks.

Left: Dingy Skipper flight sequence.

Right: Close-up of a Dingy Skipper at Llanymynech Rocks, Montgomery.

GREEN-VEINED WHITE

The Green-veined White is the UK's most common butterfly and is on the wing from April through to September. The green veins on its underwing are a visual trick created by a combination of yellow and black scales.

On a warm spring evening, I try to find roosting butterflies near where we live for the first time. At the top of our hill on Oakley Mynd is the old drovers' road. All around me lie the echoes of the past and the hills and valleys of Brockton, Bishops Castle and distant Clee Hill.

I am searching among the garlic mustard and cow parsley hoping to find a single bright emblem of new life and, finally, at the edge of a crop field, a Green-veined White settles down for the night. I kneel and get in close for a focus stack. The word 'common' is far from my mind and this swaying beauty is a distraction from the dark woes of the world, a sign that life carries on. The next day, perhaps it is the same butterfly I catch lifting from the rich blue of forget-me-nots, and, later in July, I find a mating pair glowing among the dark bracken high on the Long Mynd.

Above: This Green-veined White allowed me to crawl close in a local field.
Right: A roosting Green-veined White at dusk, south Shropshire.
Page 22: Mating Green-veined Whites, Long Mynd.
Page 23: Green-veined White in flight, Oakley Mynd.

APRIL

SPECKLED WOOD

The Speckled Wood was originally named the Wood Argus, after the mythical shepherd who had 100 eyes all over his head. He was able to always have 50 eyes open to keep guard when he went to sleep – a rather useful trait when looking after his flock! This is a butterfly that is doing well, expanding its range over England, Wales and most of Scotland. It is the only UK butterfly that can overwinter as larvae or pupae, which means it can emerge from April right through to September and occasionally October. Though it is often found in woods, I have come across it in all sorts of spots, including on one of the high paths above Old Barmouth – a wonderful sign of new spring. Normally very flighty, one individual allows me to sit right next to it and capture its bejewelled flight in the grass and the slight metallic sheen of the hairs on the back of the abdomen.

For the later emergence in August, I am walking through the delightfully wooded paths of Wenlock Edge. The light is dappled and the Speckled Woods are plentiful. I am amazed that I can get almost within touching distance of a perching individual. My macro lens reveals that even the eye is sprouting hairs. Various theories have been put forward – maybe it helps them avoid getting the eye wet when a butterfly is digging deep into dung, or could it be a way of stopping pollen sticking to the actual eye? Whatever the reason, it once again shows that the complexity of nature is a marvel and that butterflies have been hugely successful in adapting to their various habitats.

Above: A Speckled Wood in flight, Old Barmouth, Gwynedd.
Left: The eye of a Speckled Wood, Wenlock Edge, Shropshire.

SMALL COPPER

I am lying down on the path high above Old Barmouth on a sunny April morning. I really understand why butterflies hunker down on stony surfaces, as the temperature difference between the surrounding gorse bushes and the path is quite something. I put my hand down and the track actually feels warm – exactly what the Small Copper likes. If any ramblers were to make their way past, I might put them in mind of a hunter carefully stalking some big beast, but big is not a description that applies to this lustrous gemstone. My aim is to have my lens resting on the ground, level with the butterfly, and moving slowly is the recipe for not scaring it off. Then, when I am a couple of feet away, I wait and generally take a couple of thousand out-of-focus shots. What I like about butterflies and particular hotspots is that although they fly off frequently, they tend to return to the same area over and over. I finally have my moment as glitter takes to the air and 1/6400 of a second freezes motion to show aerial elegance.

On a good year, the Small Copper can have up to four broods, which means it can be on the wing even until November. In early October, chasing sightings of some of the last butterflies takes me to Bridgnorth Cemetery, where among the late-flowering ragwort and heather the final brood of Small Copper is optimistically nectaring. Among death and sleep and stone, they are a bright reminder that spring will come again.

Above: In our garden, a frequently seen form of the Small Copper with sliver-blue spots on the wings – f. *caeruleopunctata*.
Right: Small Copper take-off sequence, Old Barmouth, Gwynedd.
Pages 28-29: Small Copper at Bridgnorth Cemetery, Shropshire.

APRIL

PEARL-BORDERED FRITILLARY

Back in the 18th century, James Petiver and John Ray named both the Pearl-bordered Fritillary and the Small Pearl-bordered Fritillary the 'April Fritillary'. At that time we followed the Julian calendar, which had the new year starting 11 days earlier – their 1st May was our 19th April. With global warming and earlier emergence, this butterfly could now accurately be called the April Fritillary once again. Right at the end of the month, we head out to the Wyre Forest and in particular the central ride, which follows an old railway track and is much beloved of dog walkers, cyclists and riders.

It also contains a good number of lesser-spotting butterfly geeks and what they are all paying close attention to is the steep right-hand slope now dotted with small clumps of purple among the grass. Bugle is a perfect word for a flower that trumpets the arrival of spring, and it's delightful to see three fritillaries fighting over a single flower.

Left: Pearl-bordered Fritillary in flight above bugle flower, Wyre Forest, Worcestershire.

The Pearl-bordered always comes out before its smaller cousin. I love the patterning on its wings, which fits with the genus this butterfly sits within – *Bolos*, from the Greek for a fishing net. Today, and then into May when the bluebells come out, it is such a joy to chase a sweet slice of tangerine tint. Here in the forest and at nearby Blackgraves Copse, conservationists have been hard at work to create suitable habitat and their efforts are sorely needed, as this is one of the most threatened butterflies in the UK, with over 95% of colonies lost since 1976. I would be so saddened to see such a perfect pearl necklace broken up and scattered to the winds of uncaring history.

Left: Three Pearl-bordered Fritillaries in territorial combat over a bugle flower, Wyre Forest, Worcestershire.

SMALL HEATH

It is not even May and the Small Heaths are on the wing at Rodborough Common. This dainty, fluttery butterfly is not confined to heathland but can be seen in many different habitats. When perching, its wings are closed and the striking eye-spot is visible as a deterrent to predators. Although widespread all over the UK, it is in decline and a species of conservation concern. Today, the sun is out and I am visiting the Cotswolds, mainly on the hunt for Duke of Burgundy, but I can't complain when I not only catch the Small Heath taking off, but also its spectacular open wings, a rich yolk-yellow scrambling up from the ground.

The Small Heath can have up to three generations, and later on in August I find one roosting at The Bog mine workings, where I manage to line up sunset and silhouette in a single frame.

Above: A pair of Small Heaths, The Bog, Stiperstones, Shropshire.
Right: It's rare to see the vibrant upper wings of a Small Heath and this flight sequence shows what beauties they are, Rodborough Common, Gloucestershire.

DUKE OF BURGUNDY

No one is quite sure where the name of this butterfly arises, whether it be the rich colours reminding one of some French wines or that it resembled an ancient coat of arms. We are on stronger ground with the Latin name, *Hamearis lucina* – 'in spring, the goddess who brings light' – a good fit for this early emerging butterfly. It will certainly do for me on Prestbury Hill, near Cheltenham, where the tiny Duke is perched low in the grass. My white trousers, as ever, get totally trashed as I wiggle nearer with a fisheye wide-angle lens. This is possibly the closest I have ever got to a butterfly, my lens practically touching the insect, which finally decides it's had enough as it leaps into the picture-perfect sky. My friend Craig, in a great aside, insists I am 'the Butterfly Whisperer'. During June, later in their season, I am at Rodborough Common in a mini-heatwave. I spend hours criss-crossing the slopes with no luck and am about to head home when I spot a small fluttering of enthusiasts deep in a stiflingly hot valley dip. Of course, that's where it would be! I love the patterning of the underwing but deplore what we humans have done, as the Duke has lost 84% of its range since 1976.

Right: A male Duke of Burgundy take-off sequence, Rodborough Common, Gloucestershire.
Left: Small Heath at sunset, The Bog, Stiperstones. Having found a roosting Small Heath, I had to lie down in the grasses and wait for just before sunset, when the sun goes caramel yellow, to get this stark silhouette.
Page 38: Duke of Burgundy wide-angle sequence, Prestbury Hill, Gloucestershire.
Page 39: A female Duke of Burgundy showing three legs. The male's forelegs are reduced in size and not used for walking.

GREEN HAIRSTREAK

I love these small iridescent butterflies. If Carl Fabergé did green, surely the million or so scales on this hairstreak would fetch a fortune at auction. Nature is valuable and evolution a miracle of May. They are the only butterfly with completely green underwings, which are always closed when they perch. But the Green Hairstreak hides a secret – in flight, the upperwings are a completely different colour, a rich, reflective milk-chocolate brown. In my experience chasing Hairstreaks all over the Midlands, they are fabulously fast in flight and therefore infuriating to capture mid-motion.

With a good forecast and some light still in the sky, the nearly full moon about to rise, I head to the Stiperstones. The bumpy track leads up from Pennerley past many properties that began life as squatters' cottages. As my 4x4 chases ghosts up the track, I notice the lower slopes heaving with caramel-yellow gorse. I park up and start wandering up the track towards the Devil's Chair. In the 19th century, according to Shropshire folklore, it was said that if an ill child was pushed through the gap in the rock called the Devil's Needle, whatever ailment they suffered from would be cured. As that particular fissure has a 20-foot vertical drop on the other side, I am not sure 'cured' would be quite right. I am suddenly aware of small green spots of vibrance among the gorse and realise that I have come across a colony of roosting Green Hairstreaks. I grab a portrait infused with colour.

The moon finally agrees to lift herself above the haze and I can see an opportunity. The sun has just set and the sky is blue. Now I need to wait for the moon to intensify, line up butterfly and moon and work the shot. The Hairstreak is stilled for the night, the engine of the sun switched off as it enters a mini-nocturnal hibernation. My only desire is to celebrate its beauty, and what better frame than the round 'O' of a silver stage? We end with three colours: green, yellow and bright off-white. I can't paint, but I can do this.

Right: Moonrise, the Devil's Chair, Stiperstones.
Left: Roosting Green Hairstreak against the moon, Stiperstones, Shropshire.

Left, clockwise from top left: Roosting at dusk, Stiperstones; closed-wing portrait, Llanymynech Rocks; flight shot, Llanymynech Rocks; rarely seen brown upper wings, Llanymynech Rocks.

MARSH FRITILLARY AND SMALL BLUE

I am off for the first of many exciting safaris with fellow butterfly devotees Dave and Craig, who provide both great expertise and great company. Today, our target is Prestbury Hill in Gloucestershire, where the Small Blue is hopefully on the wing. *Cupido minimus* has lost 44% of its localities since 1976. Prestbury Hill is hidden away at the back of Cheltenham and after some slick manoeuvres through housing estates and up steep tracks, we find ourselves on a very windy ridge, the slope of the hill looking down towards the Severn. A quick cup of tea, kit checked and ready, we head downslope into the stiff breeze. There is nothing a butterfly likes less than wind and they will do their best to hunker down. As we reach the bottom, the green world turns calm. What always fascinates me is a butterfly's ability to seek out warmth. There must be a difference of at least 10 degrees down here. Out of the weatherly bluster there is a temperature pocket, and as soon as we climb over the fence I spy my first Small Blue. *Minimus* really says it all – with a wingspan of about 13mm, this is *Lepidoptera* in miniature. What I love, as it flies, is the dusting of silver-blue studs on the upperwings. Our British wildlife is, to my mind, as beautiful and important and as much in need of care and conservation as tigers and elephants.

We head further along the valley bottom. Margin areas always hold great potential, and this tucked-away slice of slope has the woods beneath too for protection and shelter to make a micro-climate. There's a shout of 'Marsh Fritillary!', which is newly emerged. It's everything the Small Blue is not – showy, loud, large and impossibly orange.

In 1884, a large swarm of their larvae was reported at marshy ground near the Ragleth above Church Stretton. Though it is long extinct in my home county, today butterfly gods have 'oranged' a visual feast for us. I am even able to get close for a focus stack, with the result turning into a tessellated mosaic.

Above and left: Marsh Fritillary in flight, Prestbury Hill, Gloucestershire.
Pages 46-47: The scales of a Marsh Fritillary.

Small Blue flight and portrait shots, Prestbury Hill.

BROWN ARGUS

I am on the Isle of Wight at Mottistone chalk pits, and right by my feet on the steep chalky slopes, a male Brown Argus is going about its business. This is one of the rare butterflies which have adapted well to changing conditions in the UK.

The decline in its preferred chalkland has seen it evolve to feed on different plants and its distribution has increased by 115% since 1976. Unlike the Northern Brown Argus, it has black spots on its forewing. I love the milk chocolate colour of the Brown Argus, and with two broods, later in August at Grafton Wood I catch a close-up portrait, posed on Devil's-bit scabious.

Above and right: Brown Argus, Grafton Wood, Worcestershire.
Left: Brown Argus flight sequence, Isle of Wight.

GLANVILLE FRITILLARY

The Glanville Fritillary is a very rare butterfly which occurs only on the Isle of Wight and Alderney and a few spots in the south of the country. Named after Eleanor Glanville, who discovered it in Lincoln in the 18th century, it was originally called the 'Lincoln Fritillary'. Eleanor was the first female naturalist and pursued her interest in butterflies after divorcing her second husband. She did not want to leave her money to her immediate family but after her death her children overturned her will by arguing she was insane because of her hobby and they succeeded. James Petiver honoured his dear friend, who was so badly treated simply because she was a woman, by titling the butterfly in her honour.

On the first day of my trip, a guided walk is full of helpful fellow enthusiasts and when I see my first Glanville Fritillary I am overjoyed. Suddenly, the long journey is totally worth it. Over the next couple of days I make new friends and mainly hang out at Compton Bay, where the sea pinks and grasses near the cliffs are a perfect habitat for this gorgeous fritillary. I arrange to meet again with local Glanville devotees Emma and Eero to see if we can find where they roost at sunset. I am trying not to hope too much, but after some searching in the low evening light we find several settling down, including a male and female who are happy to pose for portraits. As the dusk deepens, the sky and cliffs form a rich backdrop to these sleeping beauties.

Right: Glanville Fritillary flight sequence, Compton Bay, Isle of Wight.

Roosting male and female, Compton Bay. Right, clockwise from top left: On valerian at Ventnor; a mating pair at Mottistone; flight sequence at Compton Bay; roosting male and female, Compton Bay.

LARGE SKIPPER

Moses Harris was the first to describe the characteristic flight of the Large Skipper in his seminal book *The Aurelian*, published in 1766: 'When on the wing they have a kind of skipping motion, which is effected by reason of their closing their wings so often in their passage, and whenever they settle they also always close their wings'.

At the end of May, the Isle of Wight is proving a treasure trove and it's the first time in four years photographing that I have seen this stunningly marked butterfly. As ever, I am more than happy for a portrait and to grab a quick focus stack to show all that rich orange detail. With the strong black of its distinctive scent brand, the male is a very handsome specimen all round. During June, I am able to catch a female at Bury Ditches. One of the main identification clues, apart from size, is that the Large Skipper is the only skipper with hooked antennae tips, shown well in this portrait. The Large Skipper is largely stable in population and well worth seeking out.

Above: Female Large Skipper, Bury Ditches, Shropshire.
Left: Large Skipper, Mottistone Down, Isle of Wight.

ADONIS BLUE

Adonis was a mortal youth beloved by the goddess Persephone. Such a beauty lived a short time on the earth, being gored to death by a bull. It was said in Greek myth that where his blood fell, anemone flowers sprang up and Adonis himself became a symbol of eternal handsome youth. This is my third new species at Mottistone chalk pit. We descend the warm, grassy slopes to the dug-out areas where chalk was extracted. The landscape, alongside plentiful horse vetch which the larvae feed on, is exactly the right habitat for a striking butterfly. It is the male that shows off its vibrant sheen, which had historical collectors arguing over multiple titles, from the 'Celestial Blue' to the 'Ultramarine'. But Adonis Blue will do for me today, with an overcast afternoon unable to dim the burst of brightness that flares under my camera like metal in motion. The shortness of Adonis's life is a warning, as this species has lost 43% of its range from 2005-2014 and is now of conservation concern. One of the main identifiers is that the dark veins on the upperwings extend into the white fringes of the wing margins.

Later, with a second brood in August, I am on a Greenwings day out with Terry Goble. Sometimes 10 pairs of eyes make all the difference, and one of our number spots a very obliging male settled close to the ground. It's totally pristine, showing individual scales and colour-matched hairs. Handsome indeed, and worthy of a book cover!

Right: Male Adonis flight sequence.
Left: Male Adonis Blue in flight, Mottistone Down, Isle of Wight.
Page 60: Female Adonis Blue in flight, showing underside.
Page 61: Adonis Blue close-up showing scales, Mill Hill Nature Reserve, Sussex.

CHEQUERED SKIPPER

On 8th May 1798, the Reverend Dr Charles Abbot discovered and recorded the first Chequered Skipper at Clapham Park Wood in Bedfordshire. In 1828 it was noted as 'a very local species in great plenty in several parts of Northamptonshire at the end of May.' The decrease in coppicing and the replanting of broad-leaved woodland with conifers was a major cause of its decline. By 1976 it was extinct in England, though colonies were found in Scotland. In 2019, experts from Butterfly Conservation began working alongside volunteers on a reintroduction and in 2022 the success was made public.

Fast forward to a 5am start with friends to crawl through the Friday rush hour. By 10am we have pulled up at Fineshade Wood. None of us have ever seen an English Chequered Skipper and after three hours of perusing open rides filled with bramble and dog rose, we are none the wiser. It doesn't help that we keep bumping into other enthusiasts who have photographed this elusive insect. We head back to the visitor centre for much-needed coffee. After lunch, we decide to give it one more go.

As we head into the wood, a man is walking towards us down the track with a big grin on his face. It turns out our target species is up the track and far down on the right. My heart rate goes through the roof and I half walk, half run the mile or so to the spot that is filled with a bunch of blokes and a couple of women crouched in front of a bush, cameras and binos in hand.

Left: Chequered Skipper take-off showing underwings, Fineshade Wood, Northampton.

General politeness seems to be the order of the day so each of us can get in to take our shots. It's a rare good news story, where humanity has tried to reverse the effects of more ignorant times and the ghost of the Reverend Dr Charles Abbot must surely be smiling down. We too are happy as we head home with gold burnishing our memory cards and the hope that this little skipper thrives.

Above: Chequered Skipper on bramble.
Left: Chequered Skipper in flight.

HEATH FRITILLARY

I am off to Somerset on a fine May morning in search of one of Britain's rarest and most elusive butterflies. When the Large Blue became extinct in 1979, it was a shock for those involved in conservation and many were determined that the Heath Fritillary would not go down the same route. Based on research done by Martin Warren, appropriate habitat restoration and management were put in place and this pretty little butterfly was saved. Numbers are still low, with an abundance drop of 87% since 1976, but there are now thriving colonies in Somerset, Devon, Cornwall, Kent and Essex.

After a long and winding trip along the A39 and stopping to ask the way once I hit the smaller roads, I am finally looking down on the scrubby valley of Halse Combe. Once again, I am very glad of maps and specific instructions. This butterfly, in its discrete colonies, rarely travels far, and that is to my advantage. The moment I step off the path onto a semi-shaded slope full of scratchy bramble, I find my first Heath Fritillary. *Melitaea athalia* is dark orange wonder in miniature. Thalia was the daughter of Zeus associated with music and singing and its original name of *Mellicta* means honey-lick, which is of course what butterflies do best, slurping up sweet nectar with their long proboscis. As I kneel down with my wide angle, what was once called the Morning Crescent leaps into the blue sky like a soaring note. I am humbled by such rarity, grateful for the work that has gone into making Halse Combe once again alive with fluttering.

The next morning, I park up at the top of Haddon Hill and make my way downslope. The cloud is not on my side and the overcast sky convinces me this second site will be a washout. But the slightly warmer ground at the bottom of the valley is the perfect spot for a newly emerged specimen

to perch and wait for the sun. It's an opportunity to set up my tripod and use my macro to focus closely, revealing the wonder of its overlapping scales and detailed eye.

Above: Underwing flight shot; a pair of Heath Fritillary.
Left: Heath Fritillary flight sequence, Halse Combe, Somerset.
Pages 68-69: Scales close-up, Haddon Hill, Somerset; close-up of eye.

WOOD WHITE

The old title of 'Wood Lady' still suits the male Wood White, striding elegantly along wide woodland rides. This little butterfly is fairly easy to distinguish from the other whites, fluttering almost daintily close to the ground while other whites, by far the stronger flyers, zip about with purpose too fast for any walker to keep up. The Wood White was the first rare butterfly I ever photographed, as one of the main colonies in the Midlands is at Bury Ditches, the hill that overlooks our south Shropshire village. Sadly, it has lost 89% of its range since 1976 but Bury Ditches is bucking the trend, thanks to a partnership between Forestry England who own the site and Butterfly Conservation West Midlands who now manage the rides and breeding sites. As a result, the Wood White can be seen in decent numbers. More excitingly, it has spread its range to the adjoining Colstey Wood, Walcot Wood and beyond.

In close-up, the Wood White is a marvel, its wings seemingly dusted with grey chalk as if a smudge has learned how to fly. The lack of colour does me a favour by making it stand out among the green grasses and undergrowth. The first brood flies in June, followed by a smaller brood in August, and I encounter them early in the season, venturing out against instinct and in the rain. It has taken me a while to work out which butterflies perch in places we can find them, as many species go up in the trees or hide in the depths of bush and grasses. But the Wood White is a delicate chandelier who is happy to hang from forget-me-nots as the drenching rain turns each drop to shining glass.

Above: Wood White taking off from bird's foot trefoil, August.
Right: I used my camera's inbuilt stacking function to show a soaking wet Wood White waiting for the sun to dry it out again.

Wood White – I love the startled look of
the eyes in this full-frontal flight shot.

SILVER-STUDDED BLUE

The word 'pleb' comes from the Greek *plethos*, meaning 'crowd', and a pleb was a free man, without the negative connotations it now has. *Plebejus argus* is the Silver-studded Blue and the Latin feels accurate, as this little butterfly often flies freely on good years in great numbers. At Prees Heath Common Reserve, it coincides with the flowering of bell heather. The area south of Whitchurch was historically a lowland heath. In World War I, the common became a ground for trench training, with one of their number being Norval Sinclair Marley, who later in 1945 back in Jamaica fathered the great Bob Marley.

Above: Mating Silver-studded Blue, Prees Heath, Shropshire.
Right: Emerging female, attended by ants.

It transformed again in World War II into a training airfield and was then dug up to plough for crops after the war. In 2006 it was bought by Butterfly Conservation and the massive task undertaken to restore the heath.

This habitat is now the only Midlands stronghold of the Silver-studded Blue, aptly named for the metallic jewel-like studs on its wings. Aside from such showy beauty, it's an incredible example of mutualism due to its relationship with two types of black ant where both butterfly and ant benefit. From the moment the eggs are laid by the female, the nearby ants are on duty protecting the eggs and later the larvae. Once hatched,

the larvae are often carried underground into the ants' nest and carried out at night so that they can feed and grow. The ants get a good payoff – rich, sugary secretions from what is called the newcomer's gland on the tail of the larvae. As bodyguards go, the ants are the best in the business, and I am incredibly lucky to witness a female emerging out of the nest, attended by ants to clean her up ready for life on the wing.

Above: Silver-studded Blue attended by ants.
Left: Male Silver-studded Blue in flight.
Page 76-77: Silver-studded Blue at Sunset, Upper Hollesley Common, Suffolk.

LARGE HEATH

The Large Heath is the only butterfly that is confined to boggy areas, and with much land drained for agricultural use its distribution has declined, though there are nature reserves working hard to get numbers up again.

One such is Whixall Moss, and though I have been visiting for nearly 10 years, it was only in June 2022 that I was able to see what a striking butterfly it is. Although there are similarities to the Small Heath, this is much more striking, particularly the underwing spots.

Most photographs are of the male as it patrols around while the female normally stays hidden in the boggy areas, but I am able to catch a female nectaring on their favourite food plant – cross-leaved heather. The number and size of the eye spots is very variable. In Shropshire there is a well-known aberration called ab. *cockaynei*, where the underside of the hindwings is marbled with white. Due to habitat drainage, the Large Heath has declined by 58% since 1976 and it is now a priority species for conservation.

Left: Female Large Heath, Whixall Moss, Shropshire.
Right: Male Large Heath.

SMALL PEARL-BORDERED FRITILLARY

Whichever way you approach Brook Vessons on the edge of the Stiperstones, it's a good walk over tough terrain. The slopes look down towards the plain of Shrewsbury in the distance. However, the going is tough, with some of the oldest quartzite rock on the planet shattered in ankle-twisting fragments. If you stray off the track, you are soon literally bogged down in a no-man's land of tall-standing marsh thistles and tussocks that give way to sudden squelchy sinkholes where I have several times sunk up to my hips. It's not quite jungle trekking, but when at last you alight on the sign for the Reserve, it feels like an achievement.

The Small Pearl-bordered fit perfectly into this tortuous landscape. Numbers have declined by 76% since 1976, and the colonies that remain are often quite small, as is this butterfly – a fleck of orange among the green and difficult acres. I spend half an hour searching in vain and then remember to check the bottom of the slope, which is normally a heat sink. I finally spot a single specimen basking in the weak sun. As it warms up, suddenly there is a flurry of activity and I go into full-on sweaty-chase mode. It's my fifth year photographing them here and it is reassuring to come back and see them emerging – Natural England have been doing their bit by planting more marsh violet, their larval food plant.

My aim today is flight, and I swap my telephoto for a wide-angle lens and move closer in slow motion, until my camera is less than 2cm from the butterfly. It takes some doing as this fritillary mostly flies off. But I only need it to oblige once. Then the challenge is to be still and have patience, waiting for take-off until I have an action sequence that shows both butterfly and landscape, the vast view behind and orange wonder for a foreground. It is worth my soaking wet knees, my bitten and sun-burned face and the long walk in to finally be filled with the gold of gladness in my heart.

Left: Small Pearl-bordered in flight, Brook Vessons, Shropshire.
Right: Small Pearl-bordered in flight, Latterbarrow, Cumbria.
Page 82, clockwise from top left: Underwing portrait, Brook Vessons; portrait, Barmouth; underwing flight shot, Brook Vessons; underwing flight shot, Brook Vessons.
Page 83: Wide-angle flight sequence, Brook Vessons.

SWALLOWTAIL

We butterfly enthusiasts are like bees, wiggling a little dance to each other on the Internet, which, when read right, tells where good spots are to be had to gather knowledge and nectar. If jewellery could slip off and soar, then the Norfolk nature reserve of Strumpshaw Fen is a vault that is season controlled. It is worth a day's drive with hope in my heart, worth paying a venerable butterfly tour company the necessaries to be guided to a great rarity in intense June heat that only the mad would suffer. Nothing prepares one for a first encounter. Like an act of faith, I wonder if they exist at all, or if they are myth loosed upon the world, eager for credibility. When we come upon them in the reeds I am jolted, as if bright yellow electricity is loosed among the swaying hiss of beige. However, it is the chance stop-off at the famous Doctor's Garden that magic not only happens, but flowers into the sweetest of moments. When he was alive, the doctor made a garden that butted onto the footpath and filled it with most attractive sweet peas, sweet William and poppies. He had a purpose in mind. Like an angler, he threw out his scented lure, hoping to attract the UK's largest butterfly. On this day, luck and chattiness among the assembled makes for a great vibe. Friendships as brief as a butterfly's life are formed, info and sightings swapped and general bonhomie does the rounds. And when the Swallowtails come, I am suffused with vibrance. Even those who say that this is not natural habitat can be smiled upon, for the grace of the Swallowtail is forbearing and all is forgiven. With scalloped hindwings and patterning that reminds me of piano keys, it's a tune I am deliriously happy to play for the next half an hour. I am glad of the doctor, may he rest in peace.

Right: Swallowtail flying from flag iris, Strumpshaw Fen, Norfolk.

Swallowtail in flight,
the Doctor's Garden.

MOUNTAIN RINGLET

With the goodwill of others and a map to mark my target, I pull up at the parking spot for Irton Fell in Cumbria at 10 on a sunny morning. I follow the track beyond the treeline onto the fell itself, expecting good numbers in the sunshine, but strain to find a single one. Further up, I finally spy a dark flurry very close to the ground and give chase until the butterfly vanishes in the grasses. In keeping with this you-can't-see-me attitude, this butterfly was announced for the first time in the Transactions of the Entomological Society of London in 1809. It is Britain's only Alpine species and thrives at between 400-800 metres, but due to global warming has declined by 63% since 1976. Its rich brown colouring is set off by those marvellous eye-spots and a subtle copper sheen. It's known as one of the UK's hardest butterflies to find, so my lunch on the hill with a hot cup of tea feels well deserved.

Above: The view from Irton Fell, Cumbria; portrait among the grasses.

Mountain Ringlet flight sequence.

JUNE

NORTHERN BROWN ARGUS

I head down from Irton Fell feeling hot and sweaty. Under the stone bridge at Barrow-in-Furness I grab a delightful dip in the clear waters of the river. Latterbarrow Nature Reserve, above Morecambe Bay, is supposedly a slam dunk for the Northern Brown Argus. However, my experience is the opposite as I spend two hours frantically looking until I find what must be the only on-site example of the northern English subspecies *Aricia artaxerxes* ssp. *salmacis*. Perhaps the Northern Brown Argus is only just beginning to emerge. At least this one puts on a good show for me, posing for a close-up portrait and fluttering off to display those rich and

dark brown wings. The Scottish species has two pronounced white markings on the upperwings but the northern English specimen is far more subtly marked. Back in 1795 William Lewin thought it was simply a subspecies of the Brown Argus and named it the 'Brown White Spot', which is the most unpoetic name I have come across and does this northern charmer no favours.

Above: Northern Brown Argus in flight, Latterbarrow, Cumbria.
Right: Northern Brown Argus portrait.

BLACK HAIRSTREAK AND WHITE ADMIRAL

As this mission continues, I think of the arrow, plucked, strung and flung out soaring over A roads and motorways, guided by a great shared knowledge of habit, habitat, weather and season. Today, our target is Glapthorn Cow Pasture near Northampton, as we park under a tree on this hottest day of the year so far. Here, perched among the fields, is a wood of not many acres, the stronghold of Black Hairstreak. We walk into a set of rides lined with dewberry flowering in great clouds in front of the blackthorn. I'm grateful for the dappled shade, though with no breeze it's sultry and still.

Somehow, we are the first on site apart from a couple of dog walkers, and it doesn't take long before a very tatty male is spotted, my third new species this week. Take-off is slow, the heat acting like a retardant, all life going into slow motion in a green-infused sauna. A little bit of looking around and fresh females and males present themselves for as many portraits as we like.

Extra pairs of eyes pay off as another butterfly enthusiast tells us about a recently emerged White Admiral a hundred yards down the track. I love these moments when the chase is on, when you run with beating heart and yearning soul to see new life on the wing. Our luck is in. The last White Admiral I photographed, poorly, was back in 2015 and here is a male of black and white magnificence with that spectacular orangey underwing. I change my settings for flight and hope for the best, and the best is what I am given. I am so grateful for my friends' decades of knowledge, for the banter in the car and for the care and concern for the plight of British butterflies that lie under all our actions. Here is the arrow taken from the target and slid back into its quiver as we speed home with the dream of accuracy and beauty, flight and delight.

Left: White Admiral, Glapthorn Cow Pasture, Northamptonshire.
Right: Black Hairstreak in flight.
Page 94, clockwise from top left: White Admiral, Dudmaston Wood, Shropshire; Female White Admiral; Black Hairstreak portrait; Black Hairstreak flight sequence.

Male White Admiral in flight,
Glapthorne cow pasture.

RINGLET

The striking ringlet, with its velvet brown wings set off by impressive rings, is a butterfly on the up. It was named rather ponderously by Petiver as the 'Brown Eyed Butterfly with Yellow Circles', a rather dull title set against this vibrant early summer species. Range has increased by nearly 70% and abundance by nearly 400% since 1976.

I first set eyes on one of these latte-tinted delights in the beak of a spotted flycatcher. The bird has decided that the gap above our electric meter on the stone wall of our chapel is just the place to breed and bring up its chicks. As they have got older, a mere fly will not cut it, but the juicy abdomen and thorax of a Ringlet fits the bill perfectly.

I next encounter good numbers in the meadows of Millichope Estate. I get close enough for a focus stack and a wide-angle launch among the tangle of grasses. Historically, colourful butterflies were associated with joyful emotions and dark butterflies with sadder feelings. In France, the Ringlet is called *La Tristesse*, the sorrowful one. I think they got it wrong, as the sight of a sparkling Ringlet on a dull day always lifts my spirits.

Right: Spotted Flycatcher feeding Ringlet to chicks, south Shropshire; close-up of wing, Roman Bank. Page 97: Ringlet in flight.

DARK GREEN FRITILLARY

At dawn, I am on the hunt for what was long ago called The Queen of England Fritillary. It's a fair walk along old trackways past the roofless gapes of old cottages, reminders of this former hub of mining, of chapel on Sundays and dark tunnels deep into the seams of barytes. Now, the skylark sings and a dog fox feeds on something good and dead. The Small Pearl-bordered are already over, the thistles shall not host them now. They have been and now they are not, their vibrance worn to rust. However, their substitutes easily suffice. The Dark Green Fritillary is expanding its range all over south Shropshire and Brook Vessons Nature Reserve, a boggy and scratch flush, filled with dabs of darting colour.

These strong flyers rarely settle and are named after Aglaia, one of the mythical Graces, whose name translates as 'festive radiance'. The old names vanished, and though there is evidence of dark green, the tint that gets me going is sumptuous as an orange 1960s cushion. This is vibrance writ large. Add a thistle as a launching pad and a 60-frame-per-second action sequence and the glory of the morning is mine.

Above: Dark Green Fritillary, Alun Valley, south Wales.
Right: Dark Green Fritillary in flight, Bindon Hill, Dorset.

Dark Green Fritillary flight sequence,
Brook Vessons, Shropshire.

LARGE BLUE

The Large Blue was first recorded in Britain in 1795. Even at the end of the 18th century, it was considered a rare prize and many colonies vanished due mainly to over-collecting. But its final decline was caused by habitat loss and it was extinct by 1979. In the late 1970s, Jeremy Thomas discovered that their larvae fed on the grubs of a red ant, *Myrmica sabuleti*, and it was the knowledge of this parasitic relationship that enabled their reintroduction.

There are now several sites where the Large Blue is thriving once again, one of which is Daneway Banks in Gloucestershire. On an overcast morning, we pull up in the car park. The reserve is not large and the moment we walk onto the sloping site I spot my first-ever Large Blue and am in seventh heaven. I love the strong patterning on the upperwings and to catch them in flight is a joy. Here is a rare success story and after a productive morning a celebratory feast of steak sandwiches at the Daneway Inn is in order.

Top right: Large Blue underwing flight shot, Daneway Banks, Gloucestershire.
Right: Mating pair.

Male in flight.

Large Blue flight sequence.

MARBLED WHITE

The Marbled White was once 'Our Half-Mourner'. James Petiver used 'Our' to indicate that he had both collected and named the specimen, which he found in a wood near Hampstead, at that time a small village on the outskirts of London. The passing of time has treated this butterfly well and its range has increased in recent years. In south Shropshire, there is a colony on a high meadow above Roman Bank due to an unofficial reintroduction. At sunset, I find a single Marbled White roosting high on a stem of fescue. The moon is rising a few days before full, and I manage to line up moon with butterfly. I am back before dawn the following morning, when the dewdrops have lacquered the roosting butterfly that now sparkles like a precious gem, set fast in a golden landscape.

Above: Before dawn, Roman Bank.
Left: Female Marbled White wide-angle flight sequence, Roman Bank, Shropshire.
Pages 106-107: Roosting at sunset, Roman Bank.

Marbled White covered in dew at dawn, Roman Bank.

Flight shot, Roman Bank.

Portrait, Brown Clee, Shropshire.

Roosting at sunset, Bindon Hill, Dorset.

Marbled White and the rising moon, Roman Bank.

SMALL SKIPPER AND ESSEX SKIPPER

I very occasionally get stuck when it comes to identification, particularly with the difference between the Essex Skipper and Small Skipper, and have to rely on the good graces of better experts than me to lend a helping hand. The Essex Skipper has very distinctive black antennae underside tips whereas the Small Skipper has orange-brown antennae underside tips. There are other differences but this is the most obvious. I am not alone in my difficulties, as it wasn't until 1890 that the Essex was finally declared as a completely new species. F.W. Hawes first collected it in Essex in 1888 and initially assumed it was a variety of Small Skipper until the differences were analysed. The Essex Skipper now has pride of place, being the last resident UK species to be described.

Above: Essex Skipper showing skipping motion in flight, opening and closing its wings at the cut flower garden, Stokesay Flowers.
Left: Essex Skipper, Highgate Common, Staffordshire.

Essex Skipper on lavender.

Small Skippers mating,
Brown Clee, Shropshire.

PURPLE EMPEROR

Today, I am up early in a fever of excitement, chasing rumours of a rare butterfly in Warwickshire. My target is Oversley Wood, where in recent years a new inhabitant has been unofficially introduced. Its requirements are woodland rides with plenty of sallow – the caterpillar needing goat willow and grey willow to feed. After parking up I head out and soon spot one of my butterfly mates. Apparently, there is a docile female in a bush round the corner. It's a good start and I love the rich yellow of the eye, but I am after something far more striking. Soon we are rushing off down another ride to see a group with cameras and phones making a circle round a small, rocky part of the path. I join them, only to spy a brown butterfly using its proboscis to harvest minerals from the warm stones. Its Latin name is *Apatura iris*. Iris was the goddess of the rainbow who brought nectar to feed the gods and *apatura* is Greek for 'to deceive'. I am not sure why everyone is so excited until one of the gang tells me to walk round the other side of the butterfly. It is a life-changing moment. In two steps, and with a different angle, dull brown wings transform into a bright purple sheen.

I have met the Purple Emperor and his beauty is astounding. Some of the more common colours in butterflies like red, brown and black are produced by pigment, but the huge array of overlapping scales in this butterfly make the light refract and cause iridescence. Its older name of 'His Imperial Majesty' is an understatement. The male is surprisingly obliging, as they mostly spend their time up in the tree canopy feeding on aphid nectar. When they come down, they are partial to a pile of dog faeces or a smear of rotting shrimp paste. When the *Daily Mail* later run a feature with my photos, they compare the story to 'the dress' – the one online that some people saw as black and blue and others as white and gold.

Above: Male Purple Emperors often fly around the oak canopy asserting their territory. On a trip to Knepp I catch the first-ever flight sequence that shows this behaviour. This large butterfly is fearless and happily chases birds and dragonflies.

Page 115, clockwise from top left: Purple Emperor male in flight; male portrait; male underwing; female close-up.

Male Purple Emperor, with two different views showing the purple iridescence. This purple sheen varies with the intensity of light and the angle of the wings to the sun.

MEADOW BROWN

The Meadow Brown is one of our commonest butterflies whose population is stable. It can be found in most parts of the UK, with its single generation having one of the longest flight periods, often active from June through to the end of September. Today I have trekked the arduous distance of a few yards up the lane to the field behind our neighbour's house, which has been ungrazed this year and is full of wildflowers and grasses. Rain fascinates me, especially for butterflies that roost in grasses. I come across a very richly coloured Meadow Brown hunkering down in the aptly named 'meadow grass', the seed heads providing a contrast to the sprinkling of raindrops on its wing. Later in August, on our patio, I catch one taking off from my wife's scabious. I marvel at the persistence of ageing butterflies. Damaged by encounters and escapes from birds and insects, they are still able to fly even with a large portion of their wings missing. This tatty-winged Meadow Brown shows that nature will live fully to the last moment.

Above: Meadow Brown in the rain, south Shropshire.
Left: Female Meadow Brown amongst scabious flowers on our patio.

WHITE-LETTER HAIRSTREAK

It is no exaggeration to say I have been searching for five years to photograph a White-letter Hairstreak. I am sure it's not personal, but perhaps this species has been avoiding me. When Dutch elm disease ran rampant in the UK, there was a worry that the White Letter would become extinct, as this is its sole food plant. However, disease-resistant elms have regenerated in hedgerows and the butterfly has also adapted to using wych elm.

The reason they are so difficult is that they are canopy-loving and only come down onto bramble or thistle to nectar. Thus I find myself at Woodgate Country Park on the outskirts of Birmingham on a hot July afternoon. Families are bustling around and sunbathers basking like butterflies on the grass. A fellow enthusiast with far more knowledge than me finally spots one. It's a single female which has come down to nectar – a pretty butterfly all round, especially with that elegant, curled proboscis.

White-letter Hairstreaks are in serious trouble, having lost 96% abundance since 1976, and they need all the help they can get.

Left: White-letter Hairstreak nectaring on bramble blossom.
Top right: Perching high on wych elm.
Right: Male on bramble blossom.

GRAYLING

Grayling means 'the little grey one', but to my mind this butterfly is neither grey nor little. Found in the 19th century in 'the dry stony hills around Church Stretton', this year it is present in incredible numbers at the old mine workings of The Bog under the Stiperstones.

The Grayling has an excellent trick up its sleeve. On close examination, when they perch they reveal a striking and richly patterned underwing with a loud hint of orange and yellow set off with those glorious black spots with centred white dots. But among small sharp stones, grit and grass, they are completely camouflaged.

It's taken five years to capture that leap into the air, and at 1/6400 of a second I finally catch the almost never-seen wonder of their upperwing. Returning at sunset, I have my own moment of citizen science as I discover where the Grayling flies in to roost, protected under the overhang of a lichen-covered branch.

Above: Mating pair, The Bog, Stiperstones, Shropshire.
Right: Male Grayling in flight.

Top: Roosting grayling in rowan tree, The Bog.

Above: Sunset at The Bog.

Left: Close-up of eyes and scales.

Page 124: Grayling in the dew.

LULWORTH SKIPPER

Although the Lulworth Skipper extends its range both east and west from Lulworth, I like the idea of being at the exact spot where J. C. Dale Esquire rode his horse on 15th August 1832 and discovered a completely new species of butterfly. Above Lulworth Cove, Bindon Hill in more recent times recorded over 400,000 adults emerging, though the conservation record is less clear and this is now a priority species due to losses in Europe.

It is the female I am after, with her distinctive sunray pattern on the upperwing. They are the smallest of the skippers, but I finally catch one in flight on the slopes that overlook the sea, where the grasses and shrubs are filled with Marbled Whites and Dark Green Fritillaries.

Above: Female Lulworth Skipper, Bindon Hill, Dorset.
Right: Female in flight.

HIGH BROWN FRITILLARY

An early start from Shropshire gets me to the car park of Arnside Knott at 9am. I bump into a fellow butterfly devotee who knows the exact hotspot among the bracken and heather beyond the reserve's stone wall. As the sun is starting to break through, the High Brown Fritillary will hopefully be perching and warming up. It's not yet hot, so it's possible to track them as they fly.

My excitement is tempered by the hard facts. Butterfly Conservation have called it the UK's rarest butterfly and efforts have been made on several working reintroductions. Since the end of wood coppicing, there has been a 96% decline in distribution. This once common butterfly is now confined to a few sites in North West England, parts of Devon and Exmoor and several sites in Wales. It is wonderful to see at least six on the wing in a space of 15 minutes, after which the sun makes them zip about non-stop until lunch and the long motorway journey home.

Right: Take-off sequence from juniper.
Left: During flight, the underwing of the High Brown Fritillary reveals a row of brown spots between the outer margin and the silver spangles, which differentiates it from the Dark Green Fritillary, Arnside Knott, Cumbria.
Pages 130-131: Male High Brown Fritillary amongst heather in bloom.

CHALK HILL BLUE

Why have one Chalk Hill Blue when you can have four, sheltering on a pretty woolly thistle during the rain. Nottingham Hill, near Cheltenham, is a good example of the chalk downland that this butterfly favours, along with its larval food plant, horseshoe vetch. You are more likely to come across the male as it patrols low to the ground in search of a female.

This butterfly has lost 50% of its distribution since 1976 but numbers are now on the up. On my various visits here, I have been accompanied by cloud and rain; more and more I have an instinct about butterfly behaviour and the one thing that stops them in their tracks is bad weather. They generally need the sun to fly, so an overcast and wet day is perfect for catching an intimate portrait. Each raindrop transforms into a shiny Christmas bauble as the Chalk Hill sits like a statue waiting for the good graces of the sun to return.

Top right: Mating Chalk Hill Blues, Nottingham Hill, Gloucestershire.
Right: Male in flight.
Left: Four male Chalk Hill Blues on woolly thistle.

Chalk Hill Blue in the rain.

COMMON BLUE

The word 'common' derives from the Old French *comun*, meaning 'common, general, free, open, public', and there is a theory that the Common Blue is the Blue of the commons. The more common interpretation is that this is simply the most abundant of the blue butterflies, but also to my mind one of the prettiest. I first photographed the Common Blue back in 2018 at The Bog mine workings under the Stiperstones. I had spent a couple of weeks trying to work out a way to get both butterfly and Milky Way in the same raw shot without using Photoshop or compositing two pictures together. It struck me that to capture near and far, local and cosmic in a single frame during one long exposure was the ultimate geeky challenge. I needed perfect weather,

a cloudless and moonless sky and also to find a roosting Blue that was faced in the right direction to have the stars as a backdrop. For extra kit I had a remote control, flash setup and some focus-pulling adaptations bought from the USA and mainly used in filmmaking. The result was humbling as this tiny butterfly was framed against both landscape and skyscape.

Aside from flight shots and the excitement of getting a very fresh female on the wing at Grafton Wood, I was pleased to go in at dawn to the meadows at Millichope. Here was beaded dew, a glittering Blue in the glory of sunrise which was both rare and uncommon.

Above: Male Common Blue in flight, Grafton Wood, Worcestershire.

Roosting under the Milky Way,
The Bog, Shropshire.

Female Common Blue flight sequence, Grafton Wood.

Common Blue in flight at dawn,
Millichope Meadows, Shropshire.

MARBLED WHITE EMERGENCE

I have a pot of sheep fescue with hidden treasure at its roots: a tiny Marbled White larva has just transformed overnight into a pupa. Its beige case reveals nothing except a black dot, under which an eye will soon form. Two weeks go by and the chequerboard wings are revealed through its translucent surface. Concentric rings now betray an abdomen. I carry the pot everywhere, desperate to catch the moment. When it all kicks off, I am on the landing of our old chapel, sipping tea. The pupa appears to shake furiously and I grab my camera. After all the waiting, this female scrabbles out within a few seconds, clawing her way into life. It is one of the most miraculous sights I have ever encountered, a messy unfolding and then scooting up the nearest grass stem. Meconium is pumped into the wing veins to expand them and she hangs upside down to dry out. The proboscis is split in two and curls and uncurls repeatedly before fusing together. After a couple of hours, she is ready for life on the wing. As it's a very overcast day, I wait until the morning before taking her to Clun Castle, where Marbled Whites have been increasing, and I hope she soon finds a mate. What an honour to witness!

Above: Freshly emerged female Marbled White, Clun Castle.

Female Purple Hairstreak,
Prees Heath, Shropshire.

PURPLE HAIRSTREAK

The Purple Hairstreak can be found flying around the canopy of oak trees, where it feeds on honeydew, though it comes down lower to bask in sunshine. It was first called 'The Streak' based on that jagged line of white lightning on the underwing that mimics an antenna. Hairstreaks twitch their wings, perhaps in imitation of a face to warn off any hungry birds or insects. Population and range have remained fairly stable but they are often hard to find and require some patience, which is rewarded at Prees Heath by this female shimmering among the leaves.

Above: Perching for a portrait at Prees Heath, Highgate Common, Staffordshire.
Right: Close up, the scales of the Purple Hairstreak remind me of woven rattan furniture, Prees Heath, Shropshire.

SCOTCH ARGUS

I am walking Hearthstanes Estate on the Scottish Borders with kind permission from the owner. In the meadows that run along the Menzion Brook there are plenty of Common Blues and zippy Dark Green Fritillaries, but I am here for my final UK species, the Scotch Argus. I have sprayed all over with Smidge and parked up the track at an old Victorian pumping station. Vast stands of conifer line the slopes as I begin my search in earnest. I generally never believe a species exists until I hold it in my eyesight, but today dour afternoon is against me, as are the large population of midges and clegs.

But I have underestimated their brave beauty, as when I spy one deep down in the grass, its wings are wide open as if it is saying, 'Bask and be damned!' Somehow, there is a wonderful sense of Scottish bravery and that this Argus will not be put off by the weather. I get bitten as I crawl close and grab an intimate portrait. The next day, the weather is fairer. Each time the sun peers out from behind cloud, the Scotch Argus rise up as if agreeing that such warmth is a good thing and flight is the order of the day. I seem to have hit a hotspot, an unintentional meadow where the congregation of the Scotch Argus go about their spiritual business. I get my flight shots in, and when the chorus of buzzing insects keen on eating me alive is too much I retreat down the valley to where the rivers meet. There among the big boulders that line the Tweed, deep, dark, icy cool pools await me as I dip and decide that life is good. In my later research, it turns out that this is the subspecies ssp. *caledonia* – the obvious tell being each band of orange has three spots rather than four. As I drive home the next day, I marvel at such a journey, at the honour of having a Scottish colony of Scotch Argus all to myself. The crazy amount of miles, the cramp in my legs, the overloaded motorway service stations are all worth it as I wend my way south with beauty on my hard drive.

Above: Scotch Argus flight sequence.
Left: Newly emerged male Scotch Argus, Hearthstanes, Scottish Borders.

Two male Scotch Argus,
comparing a freshly emerged
example with a worn specimen.

Scotch Argus
underside portrait.

GATEKEEPER

James Petiver, who died in 1718, was the first collector to label his specimens with common vernacular names. He also named many species himself, however, many of his titles have fluttered by into the dust of history. The 'Hedge Eye with Double Specks' is a mouthful, and language also evolves and tends to conciseness, which at last led to the Gatekeeper. This vibrant butterfly is indeed the sentinel of gates and hedgerows, a colourful companion who is distinguished by a double-dotted squash ball flattened on its wing rather than the single dot of the meadow brown. At Highgate Common in Staffordshire, the end of July is marked by the blooming of bell heather. I am suddenly traversing a purple land and could not ask for a better backdrop for this perching splash of sunshine. It may be common, but for once I am utterly taken in by the Gatekeeper's contrasting splendour.

Above: Female Gatekeepers amongst heather blossom, Highgate Common, Staffordshire.
Right: Female Gatekeeper in flight, Powys Castle, Powys.

SILVER-WASHED FRITILLARY

I like a nature reserve that is only a couple of miles away and woodland rides at Bury Ditches are the perfect spot to see Silver-washed Fritillaries in August. But be aware, the male moves like the blazes, zipping along as it tries to find anything orange-looking that might be a female – I have even seen one stop to investigate a yellow leaf. The name comes from the streaks of silver found on the underwing, and if you are lucky you can see the valezina form of the female, a rich olive colour. *Fritillus* is the Latin name for a chequered gaming board and although fritillary first referred to a flower, the pattern fits well with fritillary butterflies.

In Germany it is known as the 'Emperor's Mantle', and there is something grand about this strongly coloured butterfly. Although it is a species of conservation concern, its numbers are currently stable.

Above: Male Silver-washed Fritillary in flight.
Left: Female Silver-washed Fritillary, the darker valezina form with a dusky greenish sheen, Bury Ditches, Shropshire.

Silver-washed Fritillary flight sequence.

Tatty Silver-washed Fritillary male with a freshly emerged Peacock, Bury Ditches.

SMALL TORTOISESHELL

The colours of the Small Tortoiseshell are a late summer antidote to melancholy. The name, first printed in James Petiver's catalogue of 1699, refers to the shells made from sea turtles, not tortoises. Thin slices were used to inlay furniture and for spectacle frames and boxes. Petiver had the 'Lesser Tortoiseshell' and the 'Greater Tortoiseshell', which respectively became the Small and Large. I have the dubious fortune of arriving at Knepp in Sussex, where the Large Tortoiseshell is breeding for the first time, to be told by a bunch of middle-aged men that I had just missed it. Hours of searching in the searing heatwave leads to zero sightings and the only solution is to dive into the lake for a swim and a cooldown. So for now I am happy to make do with my very local visitor, one of which I found dead and in almost perfect condition in a spider's web. Later in August, our garden landscape makes a pleasing background to an in-flight triptych, revealing wings as rare inlays of beauty and precision.

Above: Tip of the antennae of the Small Tortoiseshell. The antennae are less than 2mm in width and covered in thousands of scales. Using a macro lens with modified extension tubes, a double extender and a Raynox adaptor, then bracketing over 100 photos, a miniature landscape is revealed.
Left: Small Tortoiseshell flight sequence in our garden.

Wide-angle flight sequence.
Right: The wing scales of a
Small Tortoiseshell.

CLOUDED YELLOW

When pictures of Clouded Yellows turn up on the main butterfly Facebook groups, my heart makes a triple jump. The Warwickshire location has been kept fairly general for good reason, as this corner of a farmer's field needs not to be trampled to death. Clouded Yellows are a migratory species coming over from the Continent on a good year. The last was 2018, when they came as far inland as Venus Pool, near Shrewsbury, where a field of lucerne and bird's-foot trefoil attracted these intrepid travellers. It's going to be another scorcher, but that doesn't stop me rising early, with a few hours' driving getting me to the corner of a farmer's field and a few other keen photographers. This snippet of red clover and bramble appears to have acted as a landing pad, a big, sweet target as the Clouded Yellows have flown inland in search of new territories and nectaring sources. There are about 10 males darting around and a lot of time is spent fruitlessly crouching down to get close flight shots, only for the butterfly to fly before I am focused. It's a war of attrition and scratched knees, but time and effort finally pay off.

I chat to a local conservationist from Warwick Butterfly Conservation who is incredibly happy about our visitors. There is not much positive to be had from global warming, but these Clouded Yellows might have already bred here – and the ones I capture in flight could well be second generation. It is good to hear of the work being done to liaise with local farmers to create good habitat. With more mild winters, these butterflies might soon be added to the list of UK resident breeders. But for now, there is nothing clouded, foggy or grey about these wings – here are small bright suns, tiny supernovas among the red clover, and welcome additions to our summer.

Above: Male Clouded Yellow flight sequence, south Warwickshire.

Right: Male in flight at Venus Pool, Shropshire.

Clouded Yellow flight sequence.
Left, clockwise from top left:
Underwing flight shot; in flight;
portrait; in flight.

BRIMSTONE

Between 1699 and 1715, the Reverend Adam Buddle collected various butterfly specimens and pressed them between the leaves of his herbarium alongside local plants. It now resides in the Natural History Museum and contains the earliest specimen of the Brimstone. Brimstone is a crystal of sulphur originally used in concoctions such as brimstone and treacle, a rather dubious purgative for children, but what the male butterfly and brimstone have in common is that intense rich yellow colour. There is also the theory that the origin of 'butterfly' comes from the buttery colour of the Brimstone's wings.

The Brimstone hibernates as an adult and emerges early in the spring, with its offspring appearing in August.

It's a delight to have so many butterflies in our garden, and the valerian is a nectaring hotspot. As the August sun beats down I catch a mid-air shot where the shadow of its proboscis falls on the underwing. As an aside, Adam Buddle had a rather important plant named after him, that butterfly haven the buddleia.

Left and above: Male Brimstone in our garden.

BROWN HAIRSTREAK

Brown Hairstreaks like to stay high in the canopy of master ash trees feeding on honeydew, only occasionally coming down to sun themselves or nectar on plants such as hemp agrimony, thistle or fleabane. On a field trip to Steyning in Sussex, our tour leader spies a pretty female, posing with wings closed in the lower foliage of an oak. I finally understand why the butterfly has its name, as a streak of lightning criss-crosses her vibrant orange underwing. After a few minutes, as the sun threatens to break through the cloudy day, she opens her wings to bask.

A few days later, I find myself back in the heart of the Midlands' Brown Hairstreak territory. Grafton Wood was one of the first dedicated butterfly reserves in the country and decades of habitat work and the passion of butterfly lovers have paid off, with the Brown Hairstreak thriving. It's also a huge reserve and on this hot day my target butterfly is elusive. Over the last four years I have never managed to get a decent pic here. I have been told that hemp agrimony is the one to look out for – when the butterfly descends and nectars, this is the spot, but among the crowds of Meadow Browns and Gatekeepers it takes three sweaty hours before a single male obliges me. I dare to break the fourth wall and see how close I can get, leaning in with my wide angle, hoping for wing flare and blue sky in a single frame. When the butterfly takes off, I am hopeful. At 120 frames per second, it takes a few seconds for the camera to save the sequence from its buffer. I step back to check out the photos and once I see the spread wings and that impossible orange set against a woody backdrop, I burst out laughing. Flight and joy – that split second when all the work pays off and the glory of this good-looker is revealed.

Above: Female, Grafton Wood, Worcestershire.
Right: Male flight sequence.
Page 166: Female, Mill Hill, Sussex.
Page 167: Female, Mill Hill.

SILVER-SPOTTED SKIPPER

This is the only skipper in the UK with distinctive white markings on its underwing, which leads to its name. The second part of its Latin name, *Hesperia comma*, refers to the easily visible white comma. I have come to Aston Rowant Nature Reserve, a steep, south-facing and flower-rich chalk grassland tucked down the edge of the M40. My first visit here is overcast. When I finally get my eye in, as they really are very small, I find a single specimen hunkered down low on one of the many little paths that criss-cross these steep slopes. It's a nice warm spot to wait out the weather and I lie down with my macro and crawl in close to show all that khaki colour and fabulous array of scales, as well as a compound eye that looks prehistoric. But the deities of flight are not with me, as all I capture on the back of the screen is a rich brown blur. However, I never give up hope and a year later I return for another visit. I am overjoyed when a Silver-spotted takes off right in front of me. The good news carries on, as improvements in the management of chalk grassland sites have seen numbers rise in recent years.

Left: Silver-spotted Skipper portraits,
Aston Rowant, Oxfordshire.
Right: Silver-spotted Skipper in flight.
Pages 170-171: Scales and eye close-up.

HOLLY BLUE

If you were to ask me what is the butterfly I have seen the least of in my four years of travels round the UK, you might be surprised at the answer and also the location. Having chased the High Brown Fritillary in Cumbria and the Chequered Skipper in Northampton, I have taken three short steps from my front door to capture a little blue wonder. *Celastrina argiolus* means 'holly tree with little argus eyes', though that description does not quite work, as the second high-summer brood of the Holly Blue lays its eggs on ivy. In all this time and with so much travelling, I have only seen it a handful of times. When it settles on my wife's eryngium, it takes me a few moments to work out what species it is. I run inside to grab my camera, hoping it will carry on nectaring. To me it is

as exciting as a leucistic red kite or a black leopard. Once again I give thanks to my wife for her excellent planting, and today the spikey flowers of the eryngium have transformed into an ecosystem in miniature. The visiting bumble bees make the perfect background to a strikingly marked female who deigns to fly off in a perfect diagonal, revealing vibrant wings and the source of the old name from 1717, 'The Blue Speckt Butterfly with Black Tips'. There must be good nectaring here, for the Holly Blue returns again and again to grab the last sweetness of summer, and if I were to get the blues then let it be this bit of speedy sky among silver-sharp shadows.

Above: Female Holly Blue in flight in our garden.
Right: Flight sequence on my wife's eryngium.

PAINTED LADY

The Painted Lady was once 'Bella Donna', named for the cosmetic application of the deadly nightshade. A distillation of its berries was applied by fashionable painted ladies, i.e. courtesans, to their eyes in order to enlarge their pupils and make them more alluring. This is one of the most attractive butterflies and it has a grand story behind it that took centuries to unravel. Why did they appear in Britain in their hundreds of thousands on good years, and where did they go in winter? Hibernation was one theory, but it was not until 2009, using a combination of some 10,000 observers and radar records, before the incredible truth was revealed. They leave the UK in the autumn, often at altitudes of up to 500 metres, using the weather to clock speeds of up to 30mph. By this time, it was finally understood that the Painted Lady has the longest migration route of any species on earth. It begins in Africa, making its way north through successive generations as far as the Arctic Circle, chasing the seasons and good feeding grounds. Their 9,000-mile round trip is double that of the more well-known Monarch. Such a great desire to continue and propagate the life cycle is admirable.

By late August, one of an English generation is fuelling up on the valerian in our garden, ready for its overseas journey. The underwing pattern puts me in mind of old stained glass and I feel honoured to have such a brief visitor to our home.

Top left: At Shingle Street in Suffolk in early July, a newly arrived specimen rests on the pebble beach and allows me to catch the moment.
Left: Raindrops on a Painted Lady, south Shropshire.
Right: Wide-angle flight sequence in our garden.
Pages 176-177: Painted Lady flight sequence.

WALL

'The Golden Marbled Butterfly with Black Eyes' has flitted away into history and I am left here high on Bury Ditches, clinging to the last of summer and willing its warmth to stay a little longer. In the chaos of forestry clearance, tree stumps list over, torn from their muddy sockets. The heather is slowly regaining a hold and standing out from the grey landscape; a solitary clump of ragwort adds some colour. I have been visiting this spot for three days and am rewarded every time with at least two Walls, who strike me as bright and defiant in their continuity late into the season. *I may be going into the dark winter lands, but not yet.* My wife is perched nearby while I stalk a butterfly that has lost 76% of its distribution since 1976. The Wall butterfly does like a wall, or a bit of warm stony ground, and I find a perfectly beige path to let this bright second-brood wonder take off in front of me and show those glorious eye-filled wings.

Right: Male Wall flight sequence.

Left: Wall portrait, Bury Ditches, Shropshire.

Page 180-181: Female Wall perching on ragwort, Grafton Wood, Worcestershire.

SMALL WHITE AND LARGE WHITE

The Small White and Large White are both known as the Cabbage White and their larvae are hugely talented at decimating my wife's cabbages until their leaves resemble a Swiss cheese. Needless to say, she has very few kind words for these species, especially when I capture them evading my wife's netting to mate directly on her cabbage leaves.

I really like the Whites, especially the way they stand out from their background. One good way to distinguish Large and Small is the black mark on the forewing edges – in the Small, the black mark is smaller and in the Large it extends down the wing. These are some of the last butterflies on the wing, the second or sometimes third brood often persisting deep into September and sometimes beyond, when their appearance grows thinner and more ragged.

Right: Female Large White in flight.
Left: Small White flying from my wife's buddleia, showing extended proboscis.

A courting pair of Small Whites on my wife's brassica, one of their food plants, on which the female also lays patches of pale-yellow eggs.

At Grafton Wood, a Small White provides contrast to a stand of Ragwort.

Female Small White in flight amongst
the flowers of purple aster.

RED ADMIRAL

In 1717, James Petiver published *Papilionum Britanniae icones, nomina, &c,* which translates as *Pictures and Names of British Butterflies*. It had over 80 butterflies, indicating the level of knowledge of the time, as many were males and females of the same species. One of the species is well described as The Admiral – 'a beautiful Fly, and eminently distinguished with a red-Last cross in the upper Wing... often seen in Gardens and Fields from the End of July till Autumn'. What we don't know for sure is whether Admiral refers to Admiralty flags or an officer's uniform. Nabokov called it 'The Admirable' and there is a lot to admire about this migrant, flying over to our shores in spring, single brood emerging in late summer.

Once again, my wife's garden serves as a local nature reserve and there is plenty of buddleia and valerian for the Red Admiral to get stuck into. Using flash and macro, I zoom in on pollen, intensely coloured scales and that hairy eye. I head up the lane out of the village and something catches my eye on the pitted tarmac. I bend down to look and am amazed to see a complete red admiral wing lying in a hollow. Perhaps a bird has had a snack and discarded the wing because of little nutritional value. I pick it up gently and slip it into my wallet between a couple of petrol receipts. Later, it lies on a sheet of white paper under my attic Velux. With all my macro equipment I want to see how close in I can get. I make a focus stack of over 100 photos and blend them together. Suddenly, on my computer screen, I am staggered to see the detail in these scales. A single butterfly can have over a million scales, hung like shingles on their exoskeleton. There is an irony here – the bright colours are evolutionary and are like an advertisement hoarding with one clear, bright message: 'Don't eat me, I am toxic!' Sadly, in this case, the scales did not do their job.

Above: Close-up portrait and take-off at Shingle St, Suffolk.
Right: Launching from buddleia in the garden.

The wing scales of a Red Admiral with a kaleidoscopic effect.

A pair of Red Admirals on my wife's buddleia.

COMMA

It is 2018 and I am suddenly interested in butterflies. My wife has nurtured a glorious garden on the tiny strip of land that abuts our old Methodist chapel. Here, with late sunshine, the colours of verbena, eryngium and wallflower warm my heart. But above, beyond and better are the butterflies. This is a late harvest and names I do not immediately know nectar and leap between much sugary attraction. I am already years into my second life as a photographer and I thought the highest and noblest calling was to capture birds in flight, to make motion into feathered sculpture, reduce speed to spectacular solidity. But what about these scaly winged species? How to predict that moment of brief levitation? It seems impossible, as if the butterfly knows the three-dimensional aspects of the air and that any direction for take-off is improbably difficult to focus on.

Yet as I train my lens on a new and vibrant intimacy, I let instinctiveness guide my actions. I want this little flutter-by to go sideways, in the invisible plane of focus. The Comma, with its perfect white grammar incised onto its underwing, leans into the greenery and trusts those perfectly evolved wings to flare into a semaphored flag. Read the resulting signals and what I have then is a split second of wonder, my first-ever butterfly in flight. I like the idea of the Comma as a pause in proceedings – with a fast enough shutter speed anything is possible – and I could not ask for a better butterfly to speed me on my way. *Polygonia c-album* is indeed angle-winged, its shape unlike any other UK butterfly. Known as a 'Hop Cat' for its hop food plant, within living memory the Comma vanished along with hop farming.

But in one of those evolutionary moments of adaptation, the larvae of the Comma decided that a feast of nettles would do nicely, thank you. Over time, the Comma has spread its range even into Scotland. To capture it flying off wallflower is a great garden honour.

Above: Perching Comma in the garden.
Right: Comma feeding on blackberries, an important food source in autumn before hibernation.
Page 196: Comma flying from rudbeckia.
Page 197: Flight sequence from wallflower in the garden.

PEACOCK HIBERNATION

An accurate description of what butterflies do in this cold, dark season is overwinter. Of the UK species, nine overwinter as eggs, 32 as caterpillars, 11 as pupae and six as adults. Our butterflies lack an in-built heating system, so they rely on the engine of the sun to make them active. Come the cold weather, the six that stay as adults are playing the evolution game. By finding a cool spot deep in a tree hollow or inside a shed or dark outbuilding, it means that come the first warm days in spring, they can emerge ready to nectar and mate. It's a great strategy to get out there quickly to reproduce but not without risk, as bats, birds and spiders are always up for an immobile snack.

At Prees Heath, I have been allowed access to the old Radar Station building and am in search of roosting butterflies. I come across a Peacock hoping to ride out the winter in a cool spot out of the wind. The underneath of the Peacock is a perfect example of crypsis, a process in which butterflies merge into their environment, imitating a leaf and remaining still.

Left: Hibernating Peacock, Prees Heath Radar Station, Shropshire.

BROWN HAIRSTREAK EGGS

This morning, the sky is still streaked with orange and my breath floats in the air as I scrape the windscreen. The day promises well, especially as my mission is proof of life on a bracing January morning. After a 60-mile drive, Grafton Flyford in Worcestershire lifts itself above the valley, the thin church spire at the centre of a small quiet village. In the car park, the Aurelians gather. It is one of my favourite words, deriving from the Latin *aurelia*, meaning chrysalis or pupa. In the 1740s the Society of Aurelians used to meet in the Swan Tavern in London – they were passionate about butterflies, but back then it meant collections and pinning, quick deaths and deep study. The modern-day butterfly enthusiast is equipped with wellies and hand lenses, GPS, cameras and gloves. Today is all about egg counting and Grafton Wood is one of the best places to find them. However, at 0.6mm diameter, it is quite a task among many acres.

The Brown Hairstreak female lays its eggs on blackthorn at around 1.5 metres in height and close to the edge of the rides that criss-cross these woods. The gang from Butterfly Conservation West Midlands have been doing this for a long time and are mines of information. It's my first hunt, so although I know what to look for, I haven't got my eye in yet. Around me I keep hearing cries of 'Got one!' Considering their size, when found they leap out – a bright white pin-head dot often in the crook of a twig.

Someone spots a clump of four eggs laced into the same spot. What a find! The last quadruple was sighted in 2012. I set up with macro lens and lighting stand to focus on a tiny landscape. At this level of magnification, even the hint of a

breeze sets the eggs rocking back and forth. After an hour, when the wind calms down, I finally have my shot. The eggs are strong spiky structures resembling land-based sea urchins and the armour is useful to protect the larvae growing within. In April they hatch out and begin the process that ends in late summer with one of our most glorious butterflies emerging. Only 6% of the eggs make it to adulthood, so we are all glad to find over a hundred today, with estimates for the whole wood of around a thousand. Here now is a promise, and I walk away from the wood with a spring in my step.

Above: Female Brown Hairstreak, Grafton Wood, Worcestershire.
Right: A quartet of Brown Hairstreak Eggs, Grafton Wood.

PEARL-BORDERED FRITILLARY AND ORANGE-TIP

Above: Mating Orange-tips, South Shropshire.

Right: Pearl-bordered Fritillary nectaring on bluebells, Wyre Forest, Worcestershire.

At long last, the winter has slunk away into the shadows. Forget-me-nots add a tint of rich blue sky to the verges and the first butterflies are out and about, parading a show of colour. Here is spring come again, and hope is on the wing. In the hedgerows above our village, a pair of mating Orange-tips is surely a sign that all shall be well.

Throughout my journey up and down the UK, I have always marvelled on my return to find beauty in the local lanes and fields. Soon, another shade of orange, the Pearl-bordered Fritillary, will be buzzing over the bluebells and pausing for the sweet season's best fuel. So it goes in a dainty, speedy, effervescent cycle, a rich inheritance we must look after for the sake of our children and our children's children. I too have been flying around our green land, seeking out reserves, gardens, forests, heaths and hillsides to briefly perch in and nectar on the sweet sights and sounds of our British butterflies. I have come away with new friends, pictures to share and concern for the fragility of our local populations. If I have one hope, it is this: that the beauty contained in these pages will make you think and see that these butterflies, every one of them, rare or common, deserve our attention so that their delightful presence continues into the future.

ANDREW FUSEK PETERS

Andrew Fusek Peters is a wildlife and landscape photographer based in Shropshire. He has been on commission for the National Trust for the last eight years on the Long Mynd and Stiperstones nature reserves. His photos are regularly published in magazines and the national papers. His books include *Hill & Dale* and *Upland* (Graffeg). Andrew is an OM SYSTEM brand ambassador. www.fusekphotos.com. Instagram: @andrewfusekpeters.

FLIGHT AND AERIAL SEQUENCES

A fellow photographer who follows my work describes these type of action shots as 'bridging the gap between still shots and video'. All my photos were taken with Olympus cameras – mainly the EM1mk3 and the OM1. Both have an incredible burst feature that can take between 60 and 120 frames per second in raw. This speed is fast enough, on the rare times I get a set in focus, to reveal the butterfly's flight across the frame. I can then layer up some of these frames to show movement through space and time in a single final shot.

Capturing butterflies in flight is a challenging process, and I have spent years now studying butterfly behaviour and trying to anticipate how they will take off when nectaring and in which direction they will go. As their flight is what they are named after, it has felt like a long, tiring, but ultimately noble pursuit to explore their mid-air grace.

ACKNOWLEDGEMENTS

Many thanks to Tim Bernhard, who cast his expert eye over text and photos and saved me from much embarrassment. Thanks to Mike Williams for much help and Brown Hairstreak; and massive thanks to Butterfly Conservation West Midlands for overall support, encouragement and help over the last four years; Eero, Emma Brown, Kay Shaw and Ian Pratt on the Isle of Wight for all things Glanville; Wayne NSL; Terry Goble for all round Sussex info, Patrick Barkham for Swallowtails and support of this project; Matt Berry and Alice Hunter at Greenwings; Neil Hulme for a great morning at Knepp; Matt Cox for High Brown Fritillary expertise; Stephen and Lucy Lewis, the wise former wardens at Prees Heath, and the knowledgeable Nigel Ball for Silver-studded Blue and ants. At Whixall, the ever-helpful Stephen Barlow, and Dave M for all his game-changing help. I want to thank Ralph Gatt at Hearthstanes for access to Scotch Argus on his land, and Charlotte Cavey-Willcox. I am grateful to my fellow OM System brand ambassador Andrew McCarthy for help with Heath Fritillary; Mark Bibby and Mark Jones for info on Mountain Ringlet and Northern Brown Argus; Richard Clifford for Chalk Hill Blue; Barney and Victoria at Stokesay Flowers for skippers; James Corton in Suffolk for Silver-studded Blue; Dominic Vacher for colour proofing expertise and last but certainly not least, Melvyn Lambert for both Small Pearl-bordered and generally supportive comments all round.

Left: Marsh Fritillary flight sequence, Strawberry Banks, Gloucestershire.
Right: Small Tortoiseshell caterpillars, south Shropshire.

BIBLIOGRAPHY

Emperors, Admirals & Chimney Sweepers, Peter Marren, Little Toller.

The Aurelian Legacy, Michael A. Salmon, Harley Books.

A Natural History of the Butterflies and Moths of Shropshire, Adrian M. Riley, Swan Hill.

In Pursuit of Butterflies, Matthew Oates, Bloomsbury.

The Butterfly Isles, Patrick Barkham, Granta Books.

UK Butterflies www.ukbutterflies.co.uk

Butterfly Conservation www.butterfly-conservation.org

The Butterflies of the British Isles, Richard South, Frederick Warne & Co.

A complete guide to British Butterflies & Moths, Paul Sterry, Andrew Cleave, Rob Read, William Collins.

INDEX

Chalk Hill Blue (*Polyommatus coridon*) pages 132-134

Dingy Skipper (*Erynnis tages*) pages 16-19

Green Hairstreak (*Callophrys rubi*) pages 40-43

Large Blue (*Phengaris arion*) pages 101-103

Adonis Blue (*Polyommatus bellargus*) pages 58-61

Chequered Skipper (*Carterocephalus palaemon*) pages 62-65

Duke of Burgundy (*Hamearis lucina*) pages 37-39

Green-veined White (*Pieris napi*) pages 20-23

Large Heath (*Coenonympha tullia*) pages 78-79

Black Hairstreak (*Satyrium pruni*) pages 92-95

Clouded Yellow (*Colias croceus*) pages 158-161

Essex Skipper (*Thymelicus lineola*) pages 110-113

Grizzled Skipper (*Pyrgus malvae*) pages 18-21

Large Skipper (*Ochlodes sylvanus*) pages 56-57

Brimstone (*Gonepteryx rhamni*) pages 162-163

Comma (*Polygonia c-album*) pages 194-197

Gatekeeper (*Pyronia tithonus*) pages 148-149

Heath Fritillary (*Melitaea athalia*) pages 66-69

Large White (*Pieris brassicae*) pages 182-187

Brown Argus (*Aricia agestis*) pages 50-51

Common Blue (*Polyommatus icarus*) pages 135-139

Glanville Fritillary (*Melitaea cinxia*) pages 52-55

High Brown Fritillary (*Fabriciana adippe*) pages 128-131

Lulworth Skipper (*Thymelicus acteon*) pages 126-127

Brown Hairstreak (*Thecla betulae*) pages 164-167, 200-201

Dark Green Fritillary (*Speyeria aglaja*) pages 98-100

Grayling (*Hipparchia Semele*) pages 122-125

Holly Blue (*Celastrina argiolus*) pages 172-173

Marbled White (*Melanargia galathea*) pages 104-109

Marbled White Emergence (*Melanargia galathea*) pages 140-141

Painted Lady (*Vanessa cardui*) pages 174-177

Ringlet (*Aphantopus hyperantus*) pages 96-97

Small Copper (*Lycaena phlaeas*) pages 26-29

Speckled Wood (*Pararge aegeria*) pages 24-25

Marsh Fritillary (*Euphydryas aurinia*) pages 44-47, 204

Peacock (*Aglais io*) pages 13-17, 198-199

Scotch Argus (*Erebia aethiops*) pages 144-147

Small Heath (*Coenonympha pamphilus*) pages 34-36

Swallowtail (*Papilio machaon britannicus*) pages 84-87

Meadow Brown (*Maniola jurtina*) pages 118-119

Pearl-bordered Fritillary (*Boloria euphrosyne*) pages 30-33, 203

Silver-spotted Skipper (*Hesperia comma*) pages 168-171

Small Pearl-bordered Fritillary (*Boloria selene*) pages 80-83

Wall (*Lasiommata megera*) pages 178-181

Mountain Ringlet (*Erebia epiphron*) pages 88-89

Purple Emperor (*Apatura iris*) pages 114-117

Silver-studded Blue (*Plebeius argus*) pages 73-77

Small Skipper (*Thymelicus sylvestris*) pages 110-113

White Admiral (*Limenitis camilla*) pages 92-95

Northern Brown Argus (*Aricia artaxerxes*) pages 90-91

Purple Hairstreak (*Favonius quercus*) pages 142-143

Silver-washed Fritillary (*Argynnis paphia*) pages 150-153

Small Tortoiseshell (*Aglais urticae*) pages 154-157, 205

White-letter Hairstreak (*Satyrium w-album*) pages 120-121

Orange-tip (*Anthocharis cardamines*) pages 10-12, 202

Red Admiral (*Vanessa atalanta*) pages 188-193

Small Blue (*Cupido minimus*) pages 44-49

Small White (*Pieris rapae*) pages 182-187

Wood White (*Leptidea sinapis*) pages 70-72

Butterfly Safari
Published in Great Britain in 2023 by Bird Eye Books,
an imprint of Graffeg Limited.

Text and photographs by Andrew Fusek Peters
copyright © 2023. Designed and produced by Graffeg
Limited copyright © 2023.

Graffeg Limited, 24 Stradey Park Business Centre,
Mwrwg Road, Llangennech, Llanelli, Carmarthenshire,
SA14 8YP, Wales, UK. www.graffeg.com.

Andrew Fusek Peters is hereby identified as the
author of this work in accordance with section 77
of the Copyright, Designs and Patents Act 1988.

A CIP Catalogue record for this book is available
from the British Library.

The publisher gratefully acknowledges the financial
support of this book by the Books Council of Wales.
www.gwales.com.

Front cover: Adonis Blue, Mill Hill, Sussex.

Back cover images, left to right: Swallowtail sequence,
the Doctor's Garden, Strumpshaw Fen; Small Pearl-
bordered Fritillary, Latterbarrow, Cumbria; female
Large Heath, Whixall Moss, Shropshire; Clouded
Yellow, Venus Pool, Shropshire.

ISBN 9781802583700

1 2 3 4 5 6 7 8 9

Right: Large Blue, Daneways Banks, Gloucestershire.